Joseph Morrison

The Computation of the Transits of Venus for the Years 1874 and 1882

And of Mercury for the Year 1878

Joseph Morrison

The Computation of the Transits of Venus for the Years 1874 and 1882
And of Mercury for the Year 1878

ISBN/EAN: 9783337187033

Printed in Europe, USA, Canada, Australia, Japan

Cover: Foto ©berggeist007 / pixelio.de

More available books at **www.hansebooks.com**

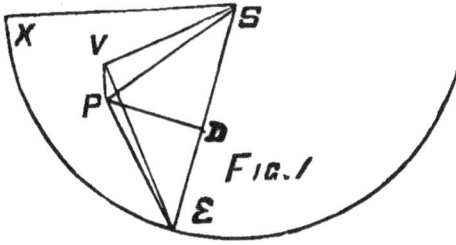

FIG. 1

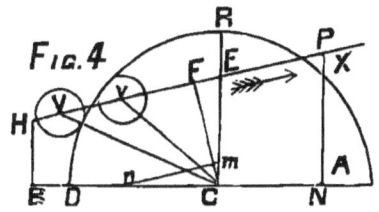

FIG. 4

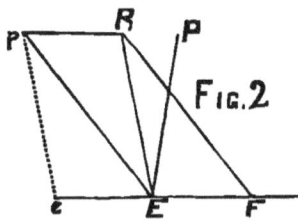

FIG. 2

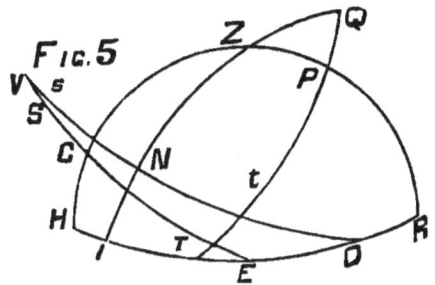

FIG. 5

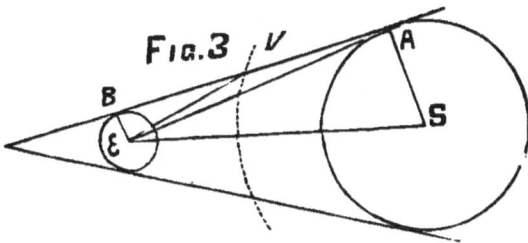

FIG. 3

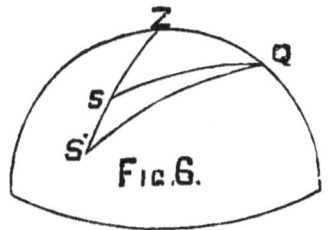

FIG. 6.

THE COMPUTATION

OF THE

TRANSITS OF VENUS

FOR THE YEARS 1874 AND 1882,

AND OF

MERCURY FOR THE YEAR 1878,

FOR THE EARTH GENERALLY AND FOR SEVERAL PLACES IN CANADA.

WITH A

POPULAR DISCUSSION OF THE SUN'S DISTANCE FROM THE EARTH, AND AN APPENDIX SHEWING THE METHOD OF COMPUTING SOLAR ECLIPSES.

BY

J. MORRISON, M. D., M. A.,

(M.B., University of Toronto),

MEMBER OF THE MEDICAL COUNCIL, AND EXAMINER IN THE COLLEGE OF PHYSICIANS AND SURGEONS OF ONTARIO.

TORONTO:
ROWSELL & HUTCHISON.
1873.

TORONTO :

PRINTED BY ROWSELL AND HUTCHISON,

KING STREET.

PREFACE.

----&----

THE following pages were drawn up for the use of Students pursuing the higher Mathematical course in our Colleges and Universities. All the necessary formulæ for calculating transits of the planets and solar eclipses from the heliocentric elements, have been investigated in order to render the work as complete in itself as possible; and while I have endeavoured to simplify the computation, I have, at the same time, given as full an account of the various circumstances attending these phenomena, as is to be found in any of the ordinary works on Spherical and Practical Astronomy.

This is, I believe, the *first* work of the kind ever published in Canada, and therefore I hope it will tend to encourage, in this country at least, the study of the grandest and noblest of the Physical Sciences.

J. M.

TORONTO, March 4th, 1873.

In preparation by the same Author.

FACTS AND FORMULÆ IN PURE AND APPLIED MATHEMATICS,

For the use of Students, Teachers, Engineers, and others.

(I.)

A TRANSIT OF VENUS.

ART. 1.—A transit of Venus over the Sun's disk, can only happen when the planet is in or near one of its nodes at the time of inferior conjunction, and its latitude, as seen from the Earth, must not exceed the sum of its apparent semi-diameter and the apparent semi-diameter of the Sun, or $31'' + 961'' = 992''$; and therefore the planet's distance from the node must not exceed $1° 50'$.

If the Earth and Venus be in conjunction at either of the nodes at any time, then, when they return to the same position again, each of them will have performed a certain number of complete revolutions.

Now the Earth revolves round the Sun in 365.256 days, and Venus in 224.7 days; and the converging fractions approximating to

$$\frac{224 \cdot 7}{365.256}, \text{ are } \frac{8}{13}, \frac{235}{382}, \frac{713}{1159}, \&c.,$$

where the numerators express the number of sidereal years, and the denominators the number of revolutions made by Venus round the Sun in the same time nearly. Therefore transits may be expected at the same node after intervals of 8 or 235 or 713 years. Now, there was a transit of Venus at the descending node, June 3rd, 1769; and one at the ascending node, December 4th, 1639. Hence, transits may be expected at the descending node in June, 2004, 2012, 2247, 2255, 2490, 2498, &c.; and at the ascending node in December, 1874, 1882, 2117, 2125, 2360, 2368, &c. In these long periods, the exact time of conjunction may differ many hours, or even four or five days from that found by the addition of the complete sidereal years, according to the preceding rule, which supposes the place of the node stationary, and that the Earth and Venus revolve round the Sun with *uniform* velocities—hypotheses which are not strictly correct. In order, therefore, to ascertain whether a transit will actually occur at these times or not, it will be necessary to calculate strictly the heliocentric longitude and latitude, and thence the geocentric longitude and latitude at the time of conjunction; then, if the geocentric latitude be less than the sum of the apparent semi-diameters of Venus and the Sun, a transit will certainly take place. The present position of Venus's nodes, is such that transits can only happen in June and December. The next four will take place December 8th, 1874, December 6th, 1882, June 7th, 2004, June 5th, 2012.

APPROXIMATE TIME OF CONJUNCTION IN LONGITUDE.

ART, 2.—From the Tables of Venus[*] and the Sun[†], we find the heliocentric longitude of the Earth and Venus to be as follows :—

Greenwich Mean Time.	Earth's Heliocen. Long.	Venus's Heliocen. Long.
Dec. 8th, 0h. (noon)	76° 17′ 33″.5	75° 52′ 55″.1
Dec. 9th, 0h. "	77° 18′ 34″.3	77° 29′ 40″.6

From which it is seen that conjunction in longitude takes place between the noons of the 8th and 9th December.

The daily motion of the Earth $= 1° 1′ 0″.8$.
The daily motion of Venus $= 1° 36′ 45″.5$.
Therefore Venus's daily gain on the Earth $= 35′ 44″.7$, and the difference of longitude of the Earth and Venus at December 8th, 0h.$=24′ 38″.4$, therefore we have .

$$35′ 44″.7 : 24′ 38″.4 :: 24h. : 16h. 32m.$$

Hence the approximate time of conjunction in longitude is December 8th, 16h. 32m.

* Tables of Venus, by G. W. Hill, Esq., of the Nautical Almanac Office, Washington, U. S.

† Solar Tables, by Hansen and Olufsen: Copenhagen, 1853. Delambre's Solar Tables. Leverrier's Solar Tables, Paris.

The exact time of conjunction will be found presently by interpolation, after we have computed from the Solar and Planetary Tables, the heliocentric places of the Earth and Venus (and thence their geocentric places) for several consecutive hours both before and after conjunction, as given below :—

Greenwich Mean Time.	Earth's Heliocentric Longitude.	Venus's Heliocentric Longitude.	Venus's Heliocentric Latitude.
Dec. 8th, 14h.	76° 53′ 8″.9	76° 49′ 21″.4	4′30″ N.
" 15h.	76 55 41 .4	76 53 23 .3	4 44 .3
" 16h.	76 58 13 .9	76 57 25 .2	4 58 .6
" 17h.	77 0 46 .5	77 1 27 .1	5 13 .
" 18h.	77 3 19 .1	77 5 29 .	5 27 .3
" 19h.	77 5 51 .7	77 6 30 .9	5 41 .6

The Sun's true longitude is found by adding 180° to the Earth's longitude.

Greenwich Mean Time.	Log. Earth's Radius Vector.	Log Venus's Radius Vector.
Dec. 8th, 14h.	9.9932897	9.8575364
" 15h.	9.9932875	9.8575336
" 16h.	9,9932854	9.8575309
" 17h.	9.9932833	9.8575281
" 18h.	9.9932811	9.8575253
" 19h.	9.0932790	9.8575225

Venus's Equatorial hor. parallax $= 33″.9 = P$. (See Art. 6.)
Sun's Equatorial hor. parallax $= 9″.1 = \pi$.
Venus's Semi-diameter $= 31″.4 = d$. (See Art. 7.)
Sun's Semi-diameter $= 16′ 16″.2 = \delta$.

The last four elements may be regarded as constant during the transit.

Sidereal time at 14h. $=$ 7h. 10m. 35.64 sec. $=$ Sun's mean longitude $+$ Nutation in $A.R.$, both expressed in time. $+ 14$ hours.

The places of Venus and the Earth, just obtained, are the *heliocentric*, or those seen from the Sun's centre. We will now investigate formulæ for computing Venus's places as seen from the Earth's centre.

ART. 3.—In *Fig.* 1, let S be the Sun's centre, E the Earth's and V that of an inferior planet, $S X$ the direction of the vernal equinox. Draw $V P$ perpendicular to the plane of the Earth's orbit, then $X S E$ is the Earth's heliocentric longitude ; $X S P$ the planet's heliocentric longitude ; $V S P$ the planet's heliocentric latitude $= l$; $V E P$ the planet's geocentric latitude $= \lambda$; $P S E$ the difference of their heliocentric longitudes, or the commutation $= C$; $P E S$ the planet's elongation $= E$; $S P E$ the planet's annual parallax $= p$; $S E$ the Earth's radius vector $= R$; $V S$ the planet's radius vector $= r$. Then in the triangle $P S E$, we have $P S = r \cos l$, $E S = R$, and angle $P S E = C$, therefore

$$R + r \cos l : R - r \cos l :: \tan \tfrac{1}{2} (p + E) : \tan \tfrac{1}{2} (p - E)$$

$$\text{But } \frac{p + E}{2} = \frac{180° - C}{2}$$

$$= 90° - \frac{C}{2}$$

Therefore $\quad 1 + \dfrac{r}{R} \cos l : 1 - \dfrac{r}{R} \cos l :: \cot \dfrac{C}{2} : \tan \tfrac{1}{2} (p - E)$

$$\text{Put } \frac{r}{R} \cos l = \tan \theta$$

Then $\qquad \tan \tfrac{1}{2} (p - E) = \dfrac{1 - \tan \theta}{1 + \tan \theta} \cot \dfrac{C}{2},$

$$= \tan (45° - \theta) \cot \frac{C}{2}, \qquad (1).$$

$$\text{and } E = 90° - \frac{C}{2} - \tfrac{1}{2} (p - E). \qquad (2).$$

Now, before conjunction, the planet will be east of the Sun, and if H be the Sun's true longitude (= the Earth's heliocentric longitude $+ 180°$), and G the geocentric longitude of the planet, we have

$$G = H \pm E \qquad (3).$$

the positive sign to be used before, and the negative sign after conjunction.

When the angle C is very small, the following method is to be preferred. Draw $P\,D$ perpendicular to $S\,E$, then

$$S\,D = r\cos l\,\cos C$$
$$P\,D = r\cos l\,\sin C,$$

Then
$$\tan E = \frac{r\cos l\,\sin C}{R - r\cos l\,\cos C}$$
$$= \frac{\tan\theta\,\sin C}{1 - \tan\theta\,\cos C}. \tag{4}$$

GEOCENTRIC LATITUDE.

ART. 4.—From the same figure we have

$$S\,P\tan l = V\,P = P\,E\tan\lambda$$

Or
$$\frac{\tan\lambda}{\tan l} = \frac{P\,S}{P\,E} = \frac{\sin E}{\sin C},$$

Therefore
$$\tan\lambda = \frac{\sin E}{\sin C}\tan l, \tag{5}$$

When the planet is in conjunction, this formula is not applicable, for then both E and C are 0°, and consequently their sines are each zero.

Since E, P and S are then in a straight line, we have

$$E\,P = R - r\cos l$$
$$\text{and } E\,P\tan\lambda = r\sin l$$

Therefore
$$\tan\lambda = \frac{r\sin l}{R - r\cos l} \tag{6}$$

DISTANCE OF THE PLANET FROM THE EARTH.

ART. 5.—
$$E\,V\sin\lambda = V\,P = r\sin l$$
$$E\,V = \frac{r\sin l}{\sin\lambda}, \tag{7}$$

When the latitudes are small the following formula is preferable :—

$$\sin E : \sin C :: P\,S : P\,E$$
$$:: r\cos l : E\,V\cos\lambda$$

From which
$$E\,V = \frac{r\sin C\,\cos l}{\sin E\,\cos\lambda}. \tag{8}$$

2

HORIZONTAL PARALLAX OF THE PLANET.

ART. 6.—Let P be the planet's horizontal parallax; π the Sun's parallax at mean distance; then, r being the planet's radius vector, expressed of course in terms of the Earth's mean distance from the Sun regarded as unity,

$$E V : 1 :: \pi : P$$

From which
$$P = \frac{\pi}{E V}$$

$$= \frac{\pi \sin \lambda}{r \sin l} \qquad (9).$$

$$= \frac{\pi}{r} \cdot \frac{\sin E \, \cos \lambda}{\sin C \, \cos l} \qquad (10).$$

APPARENT SEMI-DIAMETER OF THE PLANET.

ART. 7.—The semi-diameter of a planet, as obtained from observation with a micrometer when the planet is at a known distance, may be reduced to what it would be, if seen at the Earth's mean distance from the sun, viz., unity.

Let d' be this value of the semi-diameter, and d its value at any other time.

Then
$$E V : 1 :: d' : d$$

Therefore
$$d = \frac{d'}{E V}$$

$$= \frac{d'}{r} \cdot \frac{\sin \lambda}{\sin l}, \qquad (11).$$

$$= d' \cdot \frac{P}{\pi} \qquad (12).$$

ABERRATION IN LONGITUDE AND LATITUDE.

ART. 8.—Before computing the geocentric places of Venus by the preceding formulæ, we will first investigate formulæ for computing the aberration in longitude and latitude.

Let p and e (*Fig. 2*) be cotemporary positions of Venus and the Earth; P and E other cotemporary positions after an interval t seconds, during which time light moves from p to e or E.

If the Earth were at rest at E, Venus would be seen in the direction $p E$. Take $E F = e E$ and complete the parallelogram

ER, then pER is the aberration caused by the Earth's motion, and ep is the true direction of Venus when the earth was at e. Now RE is parallel to pe, therefore the whole aberration = PER, or the planet when at P will be seen in the direction ER.

But $PER = PEp - pER$
$$= PEp - Epe$$
= the motion of the planet round E at rest, minus the motion of E round p at rest.
= the whole geocentric motion of the planet in t seconds.

Now, light requires 8 minutes and 17.78 sec. to move from the Sun to the Earth, and if D be the planet's distance from the Earth (considering the Earth's mean distance from the Sun unity), then

$$t = D \times (8 \text{ min. } 17.78 \text{ sec.})$$
$$= 497.78 \ D.$$

And if m = the geocentric motion of the planet in one second, then

$$\text{aberration} = mt$$
$$= 497.78 \ m D. \tag{13}.$$

Resolving this along the ecliptic and perpendicular to it, we have (I being the apparent inclination of the planet's orbit to plane of the ecliptic).

$$\text{Aberration in Long.} = 497.78 \ m D \cos I \tag{14}.$$
$$\text{Aberration in Lat.} = 497.78 \ m D \sin I. \tag{15}.$$

We are now prepared to compute the apparent geocentric longitude and latitude of Venus, as well as the horizontal parallax, semi-diameter, aberration and distance from the Earth.

FOR THE GEOCENTRIC LONGITUDE.

ART. 9.—At 14 hours, we have, by using Eq. (3,) since the angle C is only 3' 47."5,

$$\log r = 9.8575364$$
$$\cos l = 9.9999996$$
$$\log R = 9.9932897$$
$$\tan \theta = 9.8642463$$
$$\theta = 36° 11' 15''$$

$$\begin{array}{ll} \tan \theta = 9.8642463, & \tan \theta = 9.8642463 \\ \cos C = 9.9999997 & \sin C = 7.0425562 \end{array}$$

$$0.731553 = \overline{9.8642460} \qquad \overline{6.9068025}$$

$$\log (1 - \tan \theta \cos C) = 9.4288569$$

$$\tan E = \overline{7.4779456}$$

$$E = 0° \; 10' \; 20''$$

Then $G = 256° \; 53' \; 8''.9 + 10' \; 20''$

$\qquad = 257 \quad 3 \quad 28 \; .9$

FOR THE GEOCENTRIC LATITUDE.

By Eq. (5).

$$\begin{array}{rl} \sin E = & 7.4779437 \\ \tan l = & 7.1169388 \\ \mathrm{cosec}\; C = & 12.9574438 \end{array}$$

$$\begin{array}{rl} \tan \lambda = & \overline{7.5523263} \\ \lambda = & 12' \; 15''.8 \text{ North.} \end{array}$$

VENUS'S DISTANCE FROM THE EARTH.

By Eq. (8.)

$$\begin{array}{rl} r = & 9.8575364 \\ \sin C = & 7.0425562 \\ \cos l = & 9.9999996 \\ \mathrm{cosec}\; E = & 12.5220563 \\ \sec \lambda = & 0.0000027 \end{array}$$

$$\log E V = \overline{9.4221512}$$

Eq. (7,) gives $\log E V = \quad 9.4221513$

VENUS'S HORIZONTAL PARALLAX.

The Equatorial Horizontal Parallax of the Sun at the Earth's mean distance will be taken $= 8''.95$, instead of $8''.577$, for reasons which will be given when we come to discuss the Sun's distance from the Earth.

By Eq. (9.)

$$\begin{array}{rl} \pi = & 0.951823 \\ \sin \lambda = & 7.552323 \end{array}$$

$$\overline{8.504146}$$

$$\begin{array}{rl} r, \text{ (ar. comp.)} = & 0.142463 \\ \mathrm{cosec}\; l = & 12.883061 \end{array}$$

$$\begin{array}{rl} \log P = & \overline{1.529670} \\ P = & 33''.9 \end{array}$$

This element is constant during the transit.

VENUS'S SEMI-DIAMETER.

Venus's semi-diameter at the Earth's mean distance from the Sun, as determined by theory and observation, is $8''.305 = d'$.

By Eq. (12.)

$$d' = 0.91934$$
$$P = 1.52967$$
$$\overline{2.44901}$$
$$\pi = 0.95182$$
$$\overline{\log d = 1.49719}$$
$$d = 31''.4, \text{ constant during transit.}$$

Some astronomers recommend the addition of about $\frac{1}{50}$ part for irradiation.

The aberration cannot be computed until we find Venus's hourly motion in orbit as seen from the Earth.

In this manner we obtain from Formulæ 1 to 12, the following results :—

Greenwich Mean Time.	Venus's Geocentric Longitude.	Venus's Geocentric Latitude.	Log. Venus's Distance from Earth.
Dec. 8th, 14h.	257° 3′ 28″.9	12′ 15″.8 N.	9.4221513
" 15h.	257 1 57 .7	12 54 .7	
" 16h.	257 0 26 .6	13 33 .7	9.4221491
" 17h.	256 58 55 .9	14 12 .9	
" 18h.	256 57 24 .8	14 ˙52 .0	9.4221342
" 19h.	256 55 54 .4	15 31 .0	

VENUS'S ABERRATION IN LONGITUDE AND LATITUDE.

ART. 10.—Venus's hourly motion in longitude is $91''$, and in latitude $39''$ (as seen from the Earth's centre). Since these are very small arcs, we may, without sensible error, regard them as the sides of a right-angled plane triangle.

Venus's hourly motion in orbit $= \sqrt{(39^2 + 91^2)} = 99''$ and therefore the motion in one second $= 0''.0275$

$$= m$$

Also $\qquad \cos I = \dfrac{91}{99} \text{ and } \sin I = \dfrac{39}{99}.$

Then by Eq. (14).

$$
\begin{aligned}
497.78 &= 2.697037 \\
m &= 8.439332 \\
D &= 9.422149 \\
\hline
&\quad 0.558518 \\
\cos I &= 9.963406 \\
\hline
\text{Aber. in long.} = 3''.32 &= 0.521924
\end{aligned}
$$

$$
\begin{aligned}
&\quad 0.558518 \\
\sin I &= 9.595429 \\
\hline
\text{Aber. in latitude} = 1''.42 &= 0.153947
\end{aligned}
$$

The aberration is constant during the transit. Since the motion of Venus is retrograde in longitude, and northward in north latitude, the aberration in longitude must be *added* to, and the aberration in latitude *subtracted* from, the planet's *true* geocentric longitude and latitude respectively in order to obtain the apparent places.

SUN'S ABERRATION.

ART. 11.—The Sun's aberration may be found from Eq. (13), by making $D = R$ and $m =$ the Sun's motion in one second.

The Sun's hourly motion in long. $= 152''.6$, and the motion in one second $= 0''.0423$

$$= m$$

Then aberration (in long.) $= 497.78 \, R \, m$

$= 20''.77$, and as the Sun always appears behind his *true* place, the aberration must be *subtracted* from the true longitude.

Applying these corrections, we obtain the following results :—

Greenwich Mean Time.	Sun's Apparent Longitude.	Venus's Apparent Geocen. Longitude.	Venus's Apparent Geocentric Latitude.
Dec. 8th, 14h.	256° 52′ 48″.2	257° 3′ 32″.2	0° 12′ 14″.4 N.
" 15h.	256 55 20 .7	257 2 01 .0	12 53 .3
" 16h.	256 57 53 .2	257 0 29 .9	13 32 .3
" 17h.	257 0 25 .8	256 58 59 .2	14 11 .5
" 18h.	257 2 58 .3	256 57 28 .1	14 50 .6
" 19h.	257 5 31 .0	256 55 57 .7	15 29 .6

APPARENT CONJUNCTION.

ART. 12.—By inspection we find that conjunction will take place between 16h. and 17h.

The relative hourly motion of the Sun and Venus is 243″.2, and the distance between them at 16h. is 156″.7.

Then   243″.2 : 156″.7 :: 1 hour : 38m. 40 sec.

During this time the Sun moves 1′ 38″.3, and Venus 58″.5 ; therefore, by collecting the elements we have :—

Greenwich M. Time of conj. in long. Dec. 8th...16h. 38m. 40sec.
Sun and Venus's longitude ....256° 59′ 31″.4.
Venus's latitude ....13′ 57″.4, N.
Venus's hourly motion in longitude....................1′ 30″.7, W.
Sun's   do.   do.   ....2′ 32″.5, E.
Venus's hourly motion in latitude..... ....39″.1, N.
Venus's horizontal parallax ....33″.9.
Sun's   do.   .... 9″.1.
Venus's semi-diameter ....31″.4.
Sun's   do.   . ...16′ 16″.2.
Obliquity of the Ecliptic......... ....23° 27′ 27″.8.
Sidereal time at 14h. (in arc).....................107° 38′ 54″.6.
Equation of time at conj. + 7m. 34 sec.

The last three elements are obtained from the Solar Tables.

TO FIND THE DURATION AND THE TIMES OF BEGINNING AND END OF THE TRANSIT FOR THE EARTH GENERALLY.

ART. 13.—The Transit will evidently commence when Venus begins to intercept the Sun's rays from the Earth, and this will take place when Venus comes in contact with the cone circumscribing the Earth and the Sun.

The semi-diameter of this cone, at the point where Venus crosses it (as seen from the centre of the Earth), is found as follows :—

Let E and S be the centres of the Earth and Sun (*Fig.* 3), and V the position of Venus at the beginning of the transit. Then the angle VES is the radius or semi-diameter of the cone where Venus crosses it.

$$VES = AES + VEA$$
$$= AES + BVE - BAE$$
$$= \delta + P - \pi \tag{16}.$$
$$= 976''.2 + 33''.9 - 9''.1 = 1001''.$$

In *Fig.* 4, take $A\,C = 1001''$; $C\,E$ at right angles to $A\,C$, $= 13'\,57''.4$; $C\,n = 4'\,03''.2$, the relative hourly motion in longitude; $C\,m = 39''.1$, the hourly motion of Venus in latitude, and through E draw $V\,X$ parallel to $m\,n$, then E is the position of Venus at conjunction, $m\,n$ is the relative hourly motion in apparent orbit, and $C\,F$ perpendicular to $V\,X$, is the least distance between their centres. The angle $E\,C\,F =$ angle $C\,n\,m$. Put $E\,C = \lambda$; $C\,n = m$; $C\,m = g$; $C\,V = C\,A$ + semi-diam. of Venus $= c$; $C\,v = C\,A -$ semi-diam. of Venus $= b$; and $T =$ the time of conjunction.

Then, by plane Trigonometry, we have $\tan n = \dfrac{g}{m}$; $m\,n = m \sec n =$ relative hourly motion in apparent orbit; $C\,F = \lambda \cos n$; $F\,E = \lambda \sin n$; time of describing $E\,F = \dfrac{\lambda}{m} \dfrac{\sin n}{\sec n}$ $= \dfrac{\lambda \sin 2n}{2\,m} = t$; therefore middle of transit occurs at $T \pm t$. (Positive sign when lat. is S.; negative when N.)

Again, $\sin V = \dfrac{\lambda \cos n}{c}$; $V\,F = c \cos V$; time of describing $V\,F = \dfrac{c}{g} \sin n \cos V = t' =$ time of describing $F\,X$, supposing the motion in orbit uniform, which it is, very nearly.

Therefore first external contact occurs at $T \pm t - t'$, and last external contact at $T \pm t + t'$.

Writing b for c, these expressions give the times of first and last internal contact.

Substituting the values of λ, c, g and m, we obtain

$$n = 9°\,7'\,33''.9.$$

Hourly motion in apparent orbit $= 246''.53$; $C\,F = 13'\,46''.8$; $E\,F = 132''.8$; time of describing $E\,F = 32\text{m.}\,19\text{sec.}$ Therefore *middle of transit* $= 16h.\,6m.\,21sec.$

Again, the angle $V = 53° 12' 41''.7$; $VF = 618''.26$, and the time of describing $VF = $ 2h. 30m. 28sec. Therefore the *first external* contact will take place at 13h. 35m. 53sec., and the *last external* contact at 18h. 36m. 49sec. The duration will therefore be 5h. 1m. very nearly.

The duration as thus determined, is not the duration of the transit as seen from the centre of the Earth, or from any point on its surface, but the whole duration from the moment Venus begins, to the moment Venus ceases to intercept the Sun's rays from *any* part of the Earth's surface.

For the time of internal contact, we have $b = 969''.6$. Then

$$\sin v = \frac{c\,F}{b},$$ or $v = 58° 30' 32''.5$; $vF = 506''.48$, and time of describing vF, 2h. 3m. 16sec. Therefore, the *first internal* contact will take place at 14h. 3m. 5sec., and the *last internal* contact at 18h. 9m. 37sec.

FROM THE EARTH'S CENTRE.

As seen from the centre of the Earth, we have at the first external contact, $c = $ the sum of their semi-diameters $= 1007''.6$, and at the first or last internal contact, $b = $ difference of their semi-diameters $= 944''.8$.

$$\text{Sin } V = \frac{FC}{c} = \frac{826.8}{1007.6},$$ therefore $V = 55° 8' 28''.5$

$VF = c \cos V = 575''.8$, and the time of describing $VF = $ 2h. 20m. 9sec. Therefore the *first external contact as seen from the Earth's centre will occur at* 13h. 46m. 12sec., *and the last external contact at* 18h. 26m. 30sec.

The *duration* $= 4$h. 40.3m.

Again, $\sin v = \dfrac{FC}{b}$, $v = 61° 3' 10''$.

$vF = b \cos v = 457''.286$, and time of describing it $=$ 1h. 51m. 17sec. Therefore,

First internal contact, 14h. 15m. 4sec.
Last internal contact, 17h. 57m. 38sec.

ART. 14.—The Sun's R. A. and Dec. are obtained from the Equations,

tan *R. A.* $=$ tan *Long.* cos *obliq.*	(17).
tan *Dec.* $=$ sin *R. A.* tan *obliq.*	(18).

3

From which we find, at conjunction,

$$\text{Sun's R. A.} = 255° \; 51' \; 53''.$$
$$= 17\text{h. } 3\text{m. } 27\text{sec.},$$
$$\text{and Sun's Dec.} = 22° \; 49' \; 15'' \text{ S.}$$

Adding 2h. 38m. 40sec. converted into sidereal time and then expressed in arc, to the sidereal time at 14h., we obtain the sidereal time at conj., $= 147° \; 25' \; 25''$. The Sun's *R. A.* at the same time $= 255° \; 51' \; 53''$, therefore the difference $108° \; 26' \; 27''$ is the Sun's distance east of Greenwich, or the east longitude of the places at which conjunction in longitude takes place at apparent noon, and that point on this meridian whose geocentric latitude is equal to the Sun's dec., will have the sun in its zenith at the same time. The Sun's dec. was found to be $22° \; 49' \; 15''$ S. $=$ the geocentric latitude which, converted into apparent or geographical latitude by Eq. (19), becomes $22° \; 57'.5$ S.

In the same way we find, that at the time of the first external contact, the Sun's *R. A.* $= 255° \; 44'$, and Dec. $22° \; 48' \; 33''$ S., and the sidereal time $= 104° \; 11'$; therefore at this time the Sun will be in the zenith of the place whose longitude is $151° \; 33'$ east (nearly), and geocentric latitude $22° \; 48' \; 33''$ S., or geographical latitude $22° \; 56' \; 50''$ S.

Similarly, we find that at the time of the *last* external contact the Sun will be in the zenith of the place whose longitude is $81° \; 23'$ E. (nearly), and geographical latitude $22° \; 58'$ S.

These points enable us to determine the places on the Earth's surface best suited for observing the transit.

TO FIND THE MOST ELIGIBLE PLACES FOR OBSERVING A TRANSIT OF VENUS.

ART. 15.—The most eligible places for observation may be determined with sufficient accuracy by means of a common terrestial globe.

From the preceding calculations, it appears that the transit will begin at 13h. 46.2m. Greenwich mean time, and continue 4h. 40.3m., and that the Sun's declination at the same time will be $22° \; 48'$ S.

Elevate the south pole 23° (nearly), and turn the globe until places in longitude $151° \; 33'$ E. are brought under the brass meridian, then the sun will be visible at the time of the first contact, at all places above the horizon of the globe, and if the globe be turned westward through 4.67 $\times$ 15° $=$ 70°, all places in the second position, will see the Sun at the time of the last contact. Those places which remain above the horizon while the globe is turned through 70° of longitude, will see the whole of the transit ; but in either position of the globe, the beginning and end of the transit will not be seen from *all* places in the horizon, but only from the points which lie in the great circle passing through the centres of Venus and the Sun.

The place which will have the Sun in the zenith at the beginning of the transit, will have the first contact on the Sun's eastern limb, and as the Sun will be near the horizon of this place when the transit ends, the duration will be diminshed by parallax.

Since Venus is in north latitude, the planet will be depressed by parallax, and consequently the duration of the transit will be diminished at all places whose south latitude is greater than the Sun's declination. For the same reason the duration will be increased at all places north of the 22nd parallel of south latitude.

Therefore from those places from which the whole transit will be visible, those which have the highest north or south latitude, should be selected, in order that the observed difference of duration may be the greatest possible.

The entire duration of this transit may be observed in eastern Siberia, Central Asia, China, and Japan. Among the most favorable southern stations, we have Australia, Tasmania, New Zealand, Auckland Island, Kerguelan's Land, and several islands in the South Pacific Ocean. For a comparison of the differences of absolute times of ingress only, or of egress only, stations differing widely both in latitude and longitude should be selected.

TO COMPUTE THE CIRCUMSTANCES OF THE TRANSIT SEEN FROM A GIVEN PLACE ON THE EARTH'S SURFACE.

ART. 16.—Before proceeding to calculate the times of beginning and end of the transit for a given place, it will be necessary to provide formulæ for computing the parallax in longitude and latitude, and in order to do this we must find :

1st. The reduction of geographical latitude due to the earth's spheroidal figure.

2nd. The reduction of the earth's equatorial radius to a given geocentric latitude, and

3rd. The altitude and (celestial) longitude of the Nonagesimal, or in other words, the distance between the poles of the ecliptic and horizon and the (celestial) longitude of the zenith of the given place at a given time.

But as this transit will not be visible in America, it will not excite that interest in this country which it otherwise would. We shall therefore omit the further consideration of it, and apply the following formulæ to the computation, for Toronto and other points in Canada, of the transit of December, 1882, which will be visible in this country.

FIRST.—REDUCTION OF LATITUDE ON THE EARTH.

ART. 17.—On account of the spheroidal figure of the Earth the meridians are ellipses, and therefore the apparent or geographical latitude does not coincide with the true or geocentric latitude, except at the equator and the poles.

Let x and y be the co-ordinates of any point on the ellipse, the origin being at the centre. The subnormal $= \dfrac{b^2}{a^2}\, x$, and if ϕ' be the geographical latitude and ϕ the geocentric.

We have
$$x \tan \phi = y$$
$$= \frac{b^2}{a^2}\, x \tan \phi'$$

Or,
$$\tan \phi = \frac{b^2}{a^2} \tan \phi'$$
$$= 0.9933254 \tan \phi' \qquad (19).$$

SECOND.—REDUCTION OF THE EARTH'S RADIUS.

ART. 18.—Let r be the radius at a place whose geocentric latitude is ϕ, x and y the co-ordinates of the place, then $x = r \cos \phi$, $y = r \sin \phi$, and by the properties of the ellipse we have $b : a :: y :$ the common ordinate on the circle described on the major axis $= \dfrac{a}{b}\, r \sin \phi$.

Therefore, $\qquad a^2 = y^2 + \dfrac{a^2}{b^2}\, r^2 \sin^2 \phi$

Or, $\qquad r^2 \cos^2 \phi + \dfrac{a^2}{b^2}\, r^2 \sin^2 \phi = a^2,$

From which $r = a \sec \phi \cos \theta,$ if $\dfrac{a}{b} \tan \phi = \tan \theta,$

or regarding a as unity, $\tan \theta = 1.003353 \tan \phi$

$(\log 1.003353 = 0.0014542),$

and $r = \sec \phi \cos \theta.$ $\qquad\qquad (20).$

The horizontal parallax of Venus obtained from Eq. (9) or (10), is the angle which the Earth's equatorial radius subtends at Venus, and is not the same for all places, but varies with the latitude.

The horizontal parallax for any place is found by multiplying the *Equatorial* horizontal parallax by the Earth's radius at that place, the equatorial radius being regarded as unity.

THIRD.—TO FIND THE ALTITUDE AND LONGITUDE OF THE NONAGESIMAL.

ART. 19.—Let $H Z R$ be a meridian, $H R$ the horizon, Z the zenith, P the pole of the equator $V E$, Q the pole of the ecliptic $V O$, V the equinox. Now since the arc joining the poles of two great circles, measures their inclination, and when produced cuts them $90°$ from their point of intersection, $N O$, $V T$, V_ζ $Q N$, each $= 90°$. Let s be the Sun's place in the ecliptic, and S his place when referred to the equator, then $V C =$ Sun's A. R. $+$ hour angle from noon $=$ sidereal time

$$= A.$$

$V N =$ longitude of the Nonagesimal $N, = m$.

$Z Q = N I$, the altitude of the Nonagesimal $= a$.

$P Q =$ the obliquity $= \omega$.

$P Z =$ co-latitude $= 90° - \phi$, (geocentric).

$\angle Z P Q = 180° - Z P T$

$\qquad\quad = 180° - (V T - V C)$

$\qquad\quad = 90° + A$, and $\angle ZQP = Nt = Vt - VN = 90° - m$

In the triangle $Z P Q$, we have

$\qquad \cos ZQ = \sin PZ \sin PQ \cos ZPQ + \cos PZ \cos PQ.$

Or, $\cos a = - \cos \phi \sin \omega \sin A + \sin \phi \cos \omega$.

Put $\sin A \cot \varphi = \tan \theta,'$

Then $\cos a = \sin \varphi \sec \theta \cos (\omega + \theta)$.   (21).

In the triangle $P\,Z\,Q$, we have

$$\sin Z\,Q : \sin Z\,P :: \sin Z\,P\,Q : \sin Z\,Q\,P$$

Or, $\sin a$: $\cos \phi$ :: $\cos A$: $\cos m$

Or, $\cos m = \cos A \cos \phi \operatorname{cosec} a$.   (22).

And from the same triangle we get

$\cos Z\,P = \sin Z\,Q \sin P\,Q \cos Z\,Q\,P + \cos Z\,Q \cos P\,Q$.

Or, $\sin \phi = \sin a \sin \omega \sin m + \cos a \cos \omega$.

From which

$$\sin m = \frac{\sin \phi - \cos a \cos \omega}{\sin a \sin \omega},$$

$$= \frac{\sin \phi - \sin \phi \cos^2 \omega + \cos \phi \sin \omega \cos \omega \sin A}{\sin a \sin \omega},$$

$$= \frac{\sin \phi \sin \omega + \cos \phi \cos \omega \sin A}{\sin a},$$

Dividing this by Equation (22), we have

$$\tan m = \frac{\tan \phi \sin \omega + \cos \omega \sin A}{\cos A},$$

$$= \tan \phi \sec A \sec \theta \sin (\omega + \theta).\quad (23).$$

Eq. (22), may now be used to find a,

$$\sin a = \cos A \cos \phi \sec m.\quad (24).$$

TO FIND THE PARALLAX IN LONGITUDE.

ART. 20.—Let Z be the zenith, Q the pole of the ecliptic, S the planet's true place, S' its apparent place, $Q\,S$ the planet's co-latitude $= 90 - \lambda$, then $Z\,Q =$ altitude of the nonagesimal $= a$, the angle $Z\,Q\,S =$ the planet's geocentric longitude — the longitude of the nonagesimal $= h$, $S\,Q\,S' =$ the parallax in longitude $= x$, and SS' is the parallax in altitude.

From the nature of parallax we have $\sin SS' = \sin P$ $\sin ZS'$ and from the triangles $S\,Q\,S'$, $Z\,Q\,S'$, we have

$$\sin x = \frac{\sin S\,S' \, \sin S'}{\sin Q\,S},$$

$$= \frac{\sin P \, \sin Z\,S' \, \sin S'}{\sin Q\,S},$$

$$= \frac{\sin P \, \sin Z\,Q \, \sin Z\,Q\,S'}{\sin Q\,S},$$

$$= \frac{\sin P \, \sin a \, \sin (h + x)}{\cos \lambda}, \qquad (25).$$

$$= k \sin (h + x), \text{ if } k = \frac{\sin P \, \sin a}{\cos \lambda};$$

and by a well known process in trigonometry,

$$x = \frac{k \sin h}{\sin 1''} + \frac{k^2 \sin 2h}{\sin 2''} + \frac{k^3 \sin 3h}{\sin 3''} + \&c. \quad (26).$$

TO FIND THE PARALLAX IN LATITUDE.

ART. 21.—In the last *Fig.* let $S'Q$ be the apparent co-latitude $= 90 - \lambda'$, then from the triangles $Q\,Z\,S$ and $Q\,Z\,S'$, we have

$$\cos Z = \frac{\cos QS - \cos QZ \, \cos ZS}{\sin QZ \sin ZS} = \frac{\cos QS' - \cos QZ \, \cos ZS'}{\sin QZ \sin ZS'}$$

or .... $$\frac{\sin \lambda - \cos a \, \cos ZS}{\sin ZS} = \frac{\sin \lambda' - \cos a \, \cos ZS'}{\sin ZS'}$$

but from the same triangles we have

$$\cos ZS = \sin a \cos \lambda \cos h + \cos a \sin \lambda$$

and $\quad \cos ZS' = \sin a \cos \lambda' \cos (h+x) + \cos a \sin \lambda'$.

which, substituted in the above, give after reduction

$$\frac{\sin ZS'}{\sin ZS} = \frac{\tan a \sin \lambda' - \cos \lambda' \, \cos (h+x)}{\tan a \sin \lambda - \cos \lambda \cos h},$$

But from the sine proportion, we have,

$$\frac{\sin ZS'}{\sin ZS} = \frac{\sin (h+x) \, \cos \lambda'}{\sin h \, \cos \lambda},$$

therefore $\dfrac{\tan a \sin \lambda' - \cos \lambda' \, \cos (h+x)}{\tan a \sin \lambda - \cos \lambda \cos h} = \dfrac{\sin (h+x) \, \cos \lambda'}{\sin h \, \cos \lambda},$

or $\quad \dfrac{\tan a \, \tan \lambda' - \cos (h + x)}{\tan a \, \tan \lambda - \cos h} = \dfrac{\sin (h + x)}{\sin h},$

From which $\tan \lambda' = \dfrac{\tan a \ \tan \lambda \sin (h + x) - \sin x}{\sin h \ \tan a}$, (27)

But $\sin x = \sin P \sin a \sec \lambda \sin (h+x).$

Therefore

$\tan \lambda' = \dfrac{\tan a \ \tan \lambda \ \sin (h+x) - \sin P \ \sin a \ \sec \lambda \ \sin (h + x)}{\sin h \ \tan a}$,

Or $\tan \lambda' = \dfrac{\sin (h + x)}{\sin h} (\tan \lambda - \sin P \cos a \sec \lambda).$

$\qquad = \dfrac{\sin (h + x)}{\sin h} (1 - \dfrac{\sin P \cos a}{\sin \lambda}) \tan \lambda.$ (28)

This formula gives the apparent latitude in terms of the true latitude and the true and apparent hour angles, but it is not in a form for logarithmic computation. We will now transform it into one which will furnish the parallax directly, and which will be adapted to logarithms.

Let $y = \lambda - \lambda'$, the parallax in latitude,

From Eq. (27) we have

$$\tan \lambda = \dfrac{\sin x}{\sin (h+x) \tan a} + \dfrac{\sin h}{\sin (h + x)} \tan \lambda'$$

Or $\tan \lambda - \tan \lambda' = \dfrac{\sin x}{\sin (h+x) \tan a} - \tan \lambda' \left(\dfrac{\sin (h+x) - \sin h}{\sin (h+x)} \right)$

Or $\dfrac{\sin (\lambda - \lambda')}{\cos \lambda \cos \lambda'} = \dfrac{\sin x}{\sin (h+x) \tan a} - \dfrac{2 \sin \frac{x}{2} \cdot \cos (h + \frac{x}{2}) \ \tan \lambda'}{\sin (h+x)}$

But $2 \sin \frac{x}{2} = \sin x \sec \frac{x}{2}$, and

$\qquad \sin x = \sin P \sin a \sec \lambda \sin (h + x)$ by Eq. (25)

Making these substitutions and reducing we have

$\sin y = \sin P \ \cos a \ \left(\cos \lambda' - \tan a \ \cos (h + \frac{x}{2}) \sec \frac{x}{2} \sin \lambda' \right)$

Put $\qquad \tan a \ \cos (h + \frac{x}{2}) \sec \frac{x}{2} = \cot \theta,$

Then $\sin y = \sin P \cos a \ \operatorname{cosec} \theta \sin (\theta - \lambda'),$

$\qquad = \sin P \cos a \ \operatorname{cosec} \theta \sin ((\theta - \lambda) + y)$ (29)

Put $\qquad \sin P \cos a \ \operatorname{cosec} \theta = k$, then as before

$y = \dfrac{k \sin (\theta - \lambda)}{\sin 1''} + \dfrac{k^2 \sin 2 (\theta - \lambda)}{\sin 2''} + \dfrac{k^3 \sin 3 (\theta - \lambda)}{\sin 3''} + \&c.$ (30)

(II.)

A TRANSIT OF VENUS,

DECEMBER 6TH, 1882.

ART. 22.—The following heliocentric positions of Venus have been computed from Hill's Tables of the Planet, and those of the Earth from Delambre's Solar Tables, partially corrected by myself, π being taken $= 8''.95$ at mean distance :—

Washington Mean Time.	Venus's Heliocentric Longitude.	Venus's Helioc. Latitude.	Log. Venus's Rad. Vector.	Earth's Helioc. Longitude.	Log. of Earth's Rad. Vector.
Dec. 5d. 21h.	74° 21' 05".88	4' 33".35 S.	9.8579538	74° 24' 46".1	9.9934344
" 22h.	74 25 07 .67	4 19 .03	9.8579510	74 27 20 5	4323
" 23h.	74 29 09 .45	4 04 .72	9.8579480	74 29 52 .9	4301
" 24h.	74 33 11 .25	3 50 .40	9.8579449	74 32 25 .4	4280
Dec. 6d. 1h.	74 37 13 .02	3 36 .09	9.8579417	74 34 57 .8	4258
" 2h	74 41 14 .78	3 21 .76	9.8579384	74 37 30 .2	4237
" 3h.	74 45 16 .56	3 07 .45	9.8579352	74 40 02 .7	9.9934215

4

Art. 23.—Passing to the true geocentric places by the aid of Formulæ (1)–(15), and then applying the correction for aberration (which, by Formulæ (14) and (15), is found to be, in longitude, + 3″.3; iu latitude + 1″.4; Sun's aberration − 20″.7), we obtain the following *apparent* geocentric places :—

Washington Mean Time.	Sun's Apparent Geocentric Longitude.	Venus's Apparent Geocentric Longitude.	Venus's Appar. Geoc. Latitude.
Dec. 5d. 21h.	254° 24′ 27″.4	254° 34′ 58″.3	12′ 28″ S.
" 22h.	26 59 .8	33 26 .7	11 49
" 23h.	29 32 .2	31 55 .2	11 10
" 24h.	32 04 .7	30 23 .6	10 30 .8
Dec. 6d. 1h.	34 37 .1	28 52 .0	9 51 .6
" 2h.	37 09 .5	27 20 .3	9 12 .5
" 3h.	39 42 .0	25 48 .6	8 33 .4

Log of Venus's distance from the Earth at noon = 9.421550. Eormulæ (9) and (12) give us $P = 33″.9$, and $d = 31″.46$, both of which may be regarded as constant during the transit.

Interpolating for the time of conjunction, and collecting the elements, we have as follows :—

Washington M. T. of Conj. in Long., Dec. 5d. 23h. 35.1m.

Venus's and Sun's longitude ....254° 31′ 01″.5

Venus's latitude .... 10′ 47″ S.

Venus's hourly motion in longitude .... 1′ 31″.6 W.

Sun's   do.   do.   .... 2′ 32″.4 E.

Venus's hourly motion in latitude .... 39″.1 N.

Sun's semi-diameter............................. 16′ 16″.2

Venus's   do.   .... 31″.5

Suns Equatorial horizontal parallax .... 9″.1

Venus's   do.   do.   .... 33″.9

Obliquity of the Ecliptic .... 23° 27′ 09″.

Sidereal time in arc at 20h..... ....195° 12′ 54″.4

Constructing a figure similar to *Fig.* 4, and employing the same notation as in Art. 13, we obtain from these elements the following results :—

$n = 9° 6′ 14″.4$; relative hourly motion in orbit, $= 247″.1$; least distance between centres, 10′ 39″;

First external contact, Dec. 5d. 20h. 50.7m. ⎫
First internal   do.,   "   21h. 11m. ⎬ Washington
Last internal   do., Dec. 6d. 2h. 48m. ⎮ Mean Time.
Last external   do.,   "   3h. 8m. ⎭

As seen from the Earth's centre.

By the formulæ of Art. 14, we find, that at the time of the *first* external contact, the Sun will be in the zenith of the place whose longitude is 45°.9 East of Washington, and latitude 22° 37′ S.; and at the last external contact the Sun will be in the zenith of the place whose longitude is 48°.3 W., and latitude 22° 41′ S.

From these data we find, by the aid of a terrestrial globe, as in the case of the transit of 1874, that the entire duration of this transit will be observed in the greater part of the Dominion of Canada, and in the United States. As Venus is south of the Sun's centre, the duration will be shortened at all places in North America, by reason of the effect of parallax. The times of first contact will be retarded at places along the Atlantic coast of Canada and the United States, while the Islands in the western part of the Indian Ocean will have this time accelerated. These localities will therefore afford good stations for determining the Sun's parallax. The time of last contact will be retarded in New South Wales, New Zealand, New Hebrides, and other Islands in the western part of the Pacific Ocean, and accelerated in the United States and the West India Islands. The duration will be lengthened in high southern latitudes, and especially in the Antarctic continent. The astronomical conditions necessary for a successful investigation of the Sun's parallax, will therefore be very favorable in this transit; and it is to be hoped that all the available resources of modern science will be employed to secure accurate observations, at all favorable points, of the times of ingress and egress of the planet on the Sun's disk, in order that we may determine with accuracy this great astronomical unit, the Sun's distance from the Earth, and thence the dimensions of the Solar System.

TO COMPUTE THE TRANSIT FOR A GIVEN PLACE ON THE EARTH'S SURFACE.

ART. 24.—Let it be required to find the times of contact for Toronto, Ontario, which is in latitude 43° 39′ 4″ N., and longitude 5h. 17m. 33sec. west of Greenwich, or 9m. 22sec. west of Washington.

Since the parallax of Venus is small, the times of ingress and egress, as seen from Toronto, will not differ much from those found for the Earth's centre. Subtracting the difference of longitude between Toronto and Washington, from the Washington Mean Time of the *first* and *last external* contacts, as given in the last article, we find the Toronto Mean Time of the first external contact to be December, 5d. 20h. 41·3m., and the last external contact to be December, 6d. 2h. 58.6m , when viewed from the centre of the earth.

The ingress will therefore occur on the east, and the egress on the west side of the meridian, and the time of ingress will consequently be retarded, and the time of egress accelerated by parallax. We therefore assume for the first external contact, December 5d. 20h. 44m., and for the last external contact, December 6d. 2h. 54m. Toronto Mean Time ; or, December 5d. 20h. 53m. 22sec , and December 6d. 3h. 3m. 22sec. Washington Mean Time.

From the elements given in Art. 23, compute for these dates the longitudes of Venus and the Sun, Venus's latitude, and the Sidereal Time in arc, at Toronto, thus :—

Washington Mean Time.	Sun's Apparent Longitude.	Venus's Appar. Longitude.	Venus's Latitude.	Sidereal Time at Toronto.
Dec. 5d. 20h 53m.22s.	254° 24′ 10″.5	254° 35′ 8″.5	12′ 32″.4 S	206° 15′ 06″
" 6d. 3h. 3m.22s.	254 39 50 .5	254 25 43 .5	8 31 .3	299 0 17

The relative positions of Venus and the Sun will be the same if we retain the Sun in his true position, and give to Venus the difference of their parallaxes, reduced to the place of observation by Art 17.

Compute next by Formulæ (19) to (30), the parallax of Venus in longitude and latitude, and apply it with its proper sign to the apparent longitude and latitude of Venus, as seen from the Earth's centre ; the results will give the planet's apparent position with respect to the Sun, when seen from the given place, and the contact of limbs will evidently happen when the apparent distance between their centres becomes equal to the sum of their semi-diameters.

We now proceed with the computation :—

By Eq. (19), $\quad \tan \phi' =\ 9.979544$

$\qquad \text{const. log} =\ 9.997091$

$\qquad\qquad \tan \phi =\ \overline{9.976635}$, therefore $\phi = 43° 27' 34''$

$\qquad \text{const. log} =\ 0.001454$

$\qquad\qquad \tan \theta =\ \overline{9.978089}$, therefore $\theta = 43° 33' 19''$

By Eq. (20), $\qquad \cos \theta =\ 9.860164$

$\qquad\qquad \sec \phi = 10.139146$

$\qquad\qquad \log r =\ \overline{9.999310}$

Diff. of Parallaxes, $24''.8 =\ 1.394452$

Reduced Parallax, $24''.76 =\ \overline{1.393762}$

ALTITUDE AND LONGITUDE OF THE NONAGESIMAL, AT THE FIRST ASSUMED TIME.

By Eq. (21),

$\sin A =\ 9.645731n \qquad\qquad \sin \phi =\ 9.837488$

$\cot \phi = 10.023366 \qquad\qquad \sec \theta = 10.042801n$

$\tan \theta =\ \overline{9.669097n} \qquad\qquad \cos (\omega + \theta) =\ \overline{9.999837n}$

$\qquad \theta = 154° 58' 42'' \qquad\qquad \cos a =\ 9.880126$

$\qquad \omega =\ 23° 27' 09'' \qquad\qquad a = 40° 38' 30''$

$\quad \omega + \theta = 178° 25' 51''$

By Eq. (23), $\qquad\qquad\qquad$ Check by Eq. (22),

$\tan \phi =\ 9.976634 \qquad\qquad \cos A =\ 9.952725n$

$\sec A = 10.047275n \qquad\qquad \cos \phi =\ 9.860854$

$\sec \theta = 10.042801n \qquad\qquad \text{cosec } a = 10.186201$

$\sin (\omega + \theta) =\ \overline{8.437493} \qquad\qquad \cos m =\ \overline{9.999780n}$

$\quad \tan m =\ \overline{8.504203} \qquad\qquad m = 181° 49' 44''$

$\qquad m = 181° 49' 44''$

PARALLAX IN LONGITUDE.

Longitude of Venus $= 254°\ 35'\ 8''.5$
Long. of the Nonagesimal $= 181°\ 49'\ 44''$
Therefore, $\quad\quad\quad\quad h = \overline{72°\ 45'\ 24''.5}$. Then by Eq. (26).

$\sin P = 6.079337$
$\sin a = 9.813799$
$\sec \lambda = \underline{10.000003}$

$k = \overline{5.893139}$	$k^2 = 1.7863$	$k^3 = 7.679$
$\sin h = 9.980029$	$\sin 2h = 9.7529$	$\sin 3h = 9.792n$
$\operatorname{cosec} 1'' = \underline{5.314425}$	$\operatorname{cosec} 2'' = \underline{5.0134}$	$\operatorname{cosec} 3'' = \underline{4.837n}$
$15''.402 = \overline{1.187593}$,	$''.0003 = \overline{4.5526}$	$= \overline{\overline{8.308n}}$

The last two terms being extremely small may be omitted, therefore the parallax in longitude $= +\ 15''.4 = x$.

PARALLAX IN LATITUDE.

By Eqs. (29) and (30).

$\tan a = 9.933672$	$\sin P = 6.079337$
$\cos (h + \tfrac{x}{2}) = 9.471860$	$\cos a = 9.880126$
$\sec \tfrac{x}{2} = \underline{10.000000}$	$\operatorname{cosec} \theta = 10.013619$
$\cot \theta = \overline{9.405532}$	$k = \overline{5.973082}$
$\theta = 75°\ 43'\ 34''.5$	$\sin (\theta + \lambda) = 9.986782$
$\lambda = 12'\ 32''.4$ S.	$\operatorname{cosec} 1'' = \underline{5.314425}$
$\theta + \lambda = 75°\ 56'\ 6''.9$.	$18''.808 = \overline{1.274289}$

$k^2 = 1.9461$	$k^3 = 7.919$
$\sin 2 (\theta + \lambda) = 9.6734$	$\sin 3 (\theta + \lambda) = 9.869n$
$\operatorname{cosec} 2'' = \underline{5.0134}$	$\operatorname{cosec} 3'' = \underline{4.837}$
$''\cdot 0004 = \overline{4.6329}$	$= \overline{8.625n}$

Therefore the parallax in latitude $= +\ 18''.8 = y$.

In the same way, we find at the second assumed time,

$$a = 27°\ 37';\ m = 317°\ 23'\ 46'';\ h = -\ 62°\ 58'\ 2''.5\ ;$$
$$x = -\ 10''.3\ ;\ y = +\ 20''.8.$$

Hence we have the following results :—

	DEC. 5D., 20H. 53M. 22SEC.		DEC. 6D., 3H. 3M. 22SEC.	
	LONGITUDE.	LATITUDE.	LONGITUDE.	LATITUDE.
Venus's Parallax.	254° 35′ 8″.5 + 15″.4	12′ 32″.4 S. +18″.8	254° 25′ 43″.5 — 10″.3	8′ 31″.3 S. + 20″.8
Sun's	254° 35′ 23″.9 254° 24′ 10″.5	12′ 51″.2	254° 25′ 33″.2 254° 39′ 50″.5	8′ 52″.1
Difference.	11′ 13″.4 Venus East.		14′ 17″.3 Venus West.	

Construct a figure similar to *Fig*. 4, make $CB = 11′ 13″.4$, and $CN = 14′ 17″.3$ the differences of longitude ; draw BH and NP below AB, because Venus is in south latitude, and make $BH = 12′ 51″.2$, and $NP = 8′ 52″.1$ the differences in latitude ; then HP will represent Venus's apparent orbit. Join HC, PC, and let V and X be the positions of the planet at the times of the first and last contacts respectively. The times of describing HV and PX are required to be found.

Proceeding in the same manner as in Art. 13, we find by plane Trigonometry, $HP = BN$ sec of the inclination of apparent orbit $= BN \ \text{✗} \ $ sec BNQ (NQ being parallel to HP)

$$\tan BNQ = \frac{BH - NP}{AC + BC}, \quad BNQ = 8° 52′ 41″ = ECF.$$

$HP = 1552″.8 = $ relative motion of Venus in 6h. 10m., therefore Venus's relative hourly motion $= 251″.8$

$$\tan BCH = \frac{BH}{BC}, \quad BCH = 48° 52′ 23″$$

$$HC = BC \ \text{sec} \ BCH = 1023″.8$$

$$HCE = 41° 7′ 37″, \ \text{hence} \ HCF = 50° 0′ 18″$$

$$CF = HC \ \cos HCF = 658″; \quad HF = HC \ \sin HCF = 784″.35$$

$$CV, \ \text{the sum of the semi-diameters} = 1007″.7$$

$$\cos VCF = \frac{CF}{CV}, \quad VCF = 49° 13′ 54″$$

$$VF = CV \ \sin VCF = 763″.19$$

$$HV = HF - VF = 21″.16.$$

Time of describing $H\,V = 5$m. 2sec., and time of describing $VF = 3$h. 1m. 51sec.

Therefore the first external contact will occur, Dec. 5d. 20h. 49m. 2sec., and the last external contact, Dec. 6d. 2h. 52m. 44sec., Mean Time at Toronto.

In a similar manner we obtain $v\,F = 677''.83$; therefore, $Vv = 85''.36$ and the time of describing $Vv = 20$m. 20sec.

Therefore the first internal contact will occur, Dec. 5d. 21h. 9m. 22sec., and the last, Dec. 6d. 2h. 32m. 24 sec., Mean Time at Toronto; or expressing these in Mean Civil Time, we have for Toronto :—

First external contact, December 6th, 8 h. 49 m., A.M.
First internal   "   "   9 h. 9·3 m., "
Last internal   "   "   2 h. 32·4 m., P.M.
Last external·   "   "   2 h. 52·7 m., "

Least distance between the centres 10′·58″.

If the highest degree of accuracy attainable be required, we must repeat the computation for the times just obtained. For ordinary purposes, however, the above times will be found sufficiently accurate.

In observing transits and solar eclipses, it is necessary to know the exact point on the Sun's disk, at which the apparent contact will take place. The angle contained by a radius drawn from the point of contact and a declination circle passing through the Sun's centre, is called the angle of position, and is computed as follows: Let LSX be a right angled spherical triangle, X the equinox, S the Sun's centre, LS a circle of latitude, perpendicular, of course, to SX, SD a declination circle; then DSX is a right angled spherical triangle, and in the present case, SD will lie to the west of SL, because the Sun's longitude lies between $180°$ and $270°$, i.e , between the autumnal equinox and the solstitial colure.

Then we have

$$\cos XS = \cot SXD \, \tan DSL.$$

Or $\tan DSL = \cos$ long $\tan \omega$.   (31).

The Sun's longitude at 8 h. 49 m., A.M., is $254° 24' 23''.2$.

Rejecting 180° we have cos 74° - 24' - 23" $= 9.429449$

$$\text{tan } \omega = 9.637317$$
$$\text{tan } DSL = 9.066766$$
$$DSL = 6° \text{ - } 39' \text{ - } 6''$$

Now the angle $VCE =$ angle $VCF -$ angle ECF
$$= 40° - 21' \, 12''$$

Therefore the angle of position is equal to the angle $DSL +$ the supplement of VCE. or $146° - 17'$. 9 from the northern limb towards the east.

In the same way we may compute the angle of position at the last external contact.

From a point in longitude 71° 55' W. of Greenwkich, and latitude 45° 21'. 7 N., at or near Bishop's College, Lennoxville, we find by the preceding method,

First external contact December 6th, 9 h. 19.5 m., A.M.
First internal   "   "   9 h. 39.4 m.,   "
Last internal   "   "   3 h. 2.6 m., P.M.
Last external   "   "   3 h. 23   m. "

Mean Time at Lennoxville.

Least distance between the centres 10' - 59".8.

From a point in longitude 64° - 24' W. of Greenwich, and latitude 45° 8' 30" N., at or near Acadia College, Wolfville, Nova Scotia.

First external contact December 6th, 9 h 48.7 m., A.M.
First internal   "   "   9 h 28.4 m.,   "
Last internal   "   "   3 h 31.7 m., P.M.
Last external   "   "   3 h 51.8 m..   "

Mean Time at Wolfville.

Least distance between the centres 10' - 59", 5.

THE SUN'S PARALLAX.

ART. 25. — A transit of Venus affords us the best means of determining with accuracy the Sun's parallax, and thence the distances of the Earth and other planets from the Sun.

5

The same things may be determined from a transit of Mercury, but not to the same degree of accuracy. The complete investigation of the methods of deducing the Sun's parallax from an observed transit of Venus or Mercury, is too refined and delicate for insertion in an elementary work like this. We add, however, the following method which is substantially the same as found in most works on Spherical Astronomy, and, which will enable the student to understand some of the general principles on which the computation depends.

TO FIND THE SUN'S PARALLAX AND DISTANCE FROM THE EARTH, FROM THE DIFFERENCE OF THE TIMES OF DURATION OF A TRANSIT OF VENUS, OBSERVED AT DIFFERENT PLACES.

ART. 26.—Let T and T' be the Greenwich mean times of the first and last contacts, as seen from the *Earth's centre;* $T+t$ and $T' + t'$ the Greenwich mean times of the first and last contacts, seen from the place of observation whose latitude is known; S and G the *true* geocentric longitudes of the Sun and Venus at the time T'; P the horizontal parallax of Venus; π the Sun's equatorial horizontal parallax; v the *relative* hourly motion of Venus and the Sun in longitude; L the geocentric latitude of Venus, and g Venus's hourly motion in latitude. Now, since Venus and the Sun are nearly coincident in position, the effect of parallax will be the same if we retain the Sun in his true position, and give to Venus the difference of their parallaxes. This difference or relative parallax is that which influences the relative positions of the two bodies.

Than $a\,(P-\pi)$, and $b\,(P-\pi)$ will be the parallax of Venus in longitude and latitude respectively, where a and b are functions of the *observed places* of Venus which depend on the observer's position on the Earth's surface. The *apparent* difference of longitude at the time T will be

$G - S + a\,(P - \pi)$; and therefore the apparent difference of longitude at the time $T + t$

$$= G - S + a\,(P - \pi) + vt,$$
and the apparent latitude of Venus at the time $T + t.$
$$= L + b\,(P - \pi) + gt.$$

Now at the time $T+t$ the distance between the centres of Venus and the Sun, is equal to the sum of their semi-diameters, $=c$, then we have

$$c^2 = \{ G - S + a\,(P - \pi) + vt \}^2 + \{ L + b\,(P' - \pi) + gt \}^2 \quad (32).$$
$$= (G - S)^2 + L^2 + 2\{ a\,(G - S) + b\,L \}\ (P' - \pi) + 2t$$
$$\{ v\,(G - S) + g\,L \};$$

neglecting the squares and products of the very small quantities t, a, b and $(P - \pi)$.

But when seen from the centre of the Earth at the time T, we have

$\quad c^2 = (G - S)^2 + L^2$, which substituted in the last equation, gives

$$t = -\frac{a\,(G - S) + b\,L}{v\,(G - S) + g\,L} \cdot (P - \pi) \quad (33).$$
$$= \delta.\,(P - \pi),\ \text{suppose}$$

Therefore the Greenwich time of the first contact at the place of observation $= T + \delta\,(P - \pi)$.

If δ' be the corresponding quantity to δ for the time T', then the time of the last contact at the place of observation

$$= T' + \delta'\,(P - \pi);$$

and if $\triangle$ be the whole duration of the transit then

$$\triangle = T' - T + (\delta' - \delta)\,(P - \pi)$$

Again, if $\triangle'$ be the duration observed at *any other* place, and β and β' corresponding values of δ and δ', we have

$$\triangle' = T' - T + (\beta' - \beta)\,(P - \pi);$$

Therefore $\quad \triangle' - \triangle = \left\{ (\beta' - \beta) - (\delta' - \delta) \right\}\,(P - \pi)$

Or, $\qquad P - \pi = \dfrac{\triangle' - \triangle}{(\beta' - \beta) - (\delta' - \delta)} \quad (34).$

Now $\qquad \dfrac{P}{\pi} = \dfrac{\text{Earth's distance from the Sun}}{\text{Earth's distance from Venus}},$

Therefore $\quad \dfrac{P - \pi}{\pi} = \dfrac{\text{Venus's distance from the Sun}}{\text{Venus's distance from the Earth}}$

$$= n,\ \text{a known quantity}$$

$$\pi = \frac{1}{n}\,(P - \pi). \qquad (35).\text{—(\textit{Hymers's Astron.})}$$

or $\qquad \pi = \dfrac{P}{n + 1}\ .$

If the first or last contact *only* be observed, the place of observation should be so selected that, at the beginning or end of the transit, the sun may be near the horizon (say 20° above it) in order that the time of beginning or end may be accelerated or retarded as much as possible by parallax.

Again, since t is known in Eq. (33), being the difference of the Greenwich mean times of beginning or end, as seen from the Earth's centre and the place of observation, we have from Eq. (32) by eliminating c,

$$(P-\pi)^2 + \frac{2a\,(G-S+vt) + 2b\,(L+gt)}{a^2 + b^2}\,(P-\pi) =$$

$$-\frac{t^2\,(v^2+g^2) + 2t\,(v\,(G-S) + Lg)}{a^2 + b^2}$$

Or, $(P-\pi)^2 + A\,(P-\pi) = B$, suppose.   (36).

And let $(P-\pi)^2 + C\,(P-\pi) = D$, be a similar equation derived from observation of the first or last contact at another place, then

$$(A-C)\,(P-\pi) = B-D$$

Or, $$P-\pi = \frac{B-D}{A-C},$$ (37).

And $$\pi = \frac{1}{n}\,(P-\pi),\text{ as before.}$$

THE SUN'S DISTANCE FROM THE EARTH.

ART. 27.—If D' represent the Sun's distance, and r the Earth's equatorial radius, then

$$D' = \frac{r}{\sin \pi}$$

$$= r\,\frac{206264 \cdot 8}{\pi}$$ (38).

From the observations made during the Transit of 1769, the Sun's equatorial horizontal parallax (π) at mean distance, was determined to be 8″.57 which, substituted in the last equation, gives for the Sun's mean distance $24068.23r$, or in round numbers 95,382,000 miles ; but recent investigations in both physical and practical astronomy, have proved beyond all doubt that this value is too great by about four millions of miles.

In determining the Solar parallax from a transit of an inferior planet, two methods are employed. The first, and by far the best, consists in the comparison of the observed duration of the transit at places favorably situated for shortening and lengthening it by the effect of parallax. This method is independent of the longitudes of the stations, but it cannot be always applied with advantage in every transit, and fails entirely when any atmospherical circumstances interfere with the observations either at the first or last contact. The other consists in a comparison of the absolute times of the *first* external or internal contact *only*, or of the *last* external or internal contact *only*, at places widely differing in latitude. The longitudes of the stations enter as essential elements, and they must be well known in order to obtain a reliable result. The transit of 1761 was observed at several places in Europe, Asia, and Africa, but the results obtained from a full discussion of the observations by different computers, were unsatisfactory, and exhibited differences which it was impossible to reconcile. That transit was not therefore of much service in the solution of what has been justly termed "the noblest problem in astronomy." The most probable value of the parallax deduced from it, was 8″.49. The partial failure was due to the fact that it was impossible to select such stations as would give the first method a fair chance of success, and as there was considerable doubt about the correctness of the longitudes of the various observers, the results obtained from the second method could not be depended on.

The unsatisfactory results obtained from the transit of 1761, gave rise to greater efforts for observing the one of 1769, and observers were sent to the Island of Tahiti, Manilla, and other points in the Pacific Ocean; to the shores of Hudson's Bay, Madras, Lapland, and to Wardhus, an Island in the Arctic Ocean, at the north-east extremity of Norway. The first external and internal contacts were observed at most of the European observatories, and the last contacts at several places in Eastern Asia and in the Pacific Ocean; while the whole duration was observed at Wardhus, and other places in the north of Europe, at Tahiti, &c. But on account of a cloudy atmosphere at all the northern stations, except Wardhus, the entire duration of the transit could not be observed, and it consequently happened that the observations taken at Wardhus exercised a great influence on the final result. This, however, would have been a matter of very little importance, if the observations taken there by the observer, Father Hell, had been reliable, but they exhibited such differences from those of other observers, as to lead some to regard them as forgeries. A careful examination of all the available observations of this transit, gave 8″.57 for the solar parallax, and consequently 95,382,000 miles for the Sun's mean distance.

The first serious doubts as to the accuracy of this value of the Solar parallax, began to be entertained in the year 1854, when Professor Hansen found from an investigation of the lunar orbit, and especially of that irregularity called the *parallactic equation* which depends on the Earth's distance from the Sun, that the Moon's place as deduced from the Greenwich observations, did not agree with that computed with the received value of the Sun's distance, which he found to require a considerable diminution. The same conclusion was confirmed by an examination of a long series of lunar observations taken at Dorpat, in Russia. The value of the solar parallax thus indicated by theory and observation, is 8″.97 which is about four-tenths of a second greater than that obtained from observations of the transit of Venus in 1769 ; and if this value of the parallax be substituted in Eq. (38), it will be found to give a diminution of more than 4,000,000 miles in the Earth's mean distance from the Sun.

A few years ago M. LeVerrier, of Paris, found, after a most laborious and rigorous investigation of the observations on the Moon, Sun, Venus, and Mars, taken at Greenwich, Paris, and other observatories, that an augmentation of the Solar parallax or a dimination of the hitherto received distance of the Earth from the Sun, to an amount almost equal to that previously assigned by Professor Hansen, was absolutely necessary to account satisfactorily for the lunar equation which required an increase of a twelfth part, and for the excessive motions of Venus's nodes, and the perihelion of Mars. He adopted 8″.95 for the Solar parallax.

The most recent determination of the velocity of light combined with the time which it requires to travel from the Sun to the Earth, viz.: 8 minutes and 18 seconds very nearly, affords another independent proof that the commonly received distance is too great by about $\frac{1}{30}$th part. The value of the Solar parallax indicated by this method is $8''.86$.

The great eccentricity of the orbit of Mars causes a considerable variation in the distance of this planet from the Earth at the time of opposition. Sometimes its distance from the Earth is only a little more than one-third of the Earth's distance from the Sun. Now, if Mars when thus favorably situated, be observed on the meridians of places widely differing in latitude—such as Dorpat and the Cape of Good Hope—and if the observations be reduced to the same instant by means of the known velocity of the planet, we shall, after correcting for refraction and instrumental errors, possess data for determining with a high degree of accuracy, the planet's distance from the Earth, and thence the Sun's distance and parallax. The oppositions of 1860 and 1862, were very favorable for such observations, and attempts were made at Greenwich, Poulkova, Berlin, the Cape of Good Hope, Williamstown, and Victoria, to determine the Solar parallax at those times. The mean result obtained from these observations, was $8''.95$ which agrees exactly with the theoretical value of the parallax previously obtained by M. LeVerrier.

Hence, we find that a diminution in the Sun's distance, as commonly received, is indicated, 1st, By the investigation of the parallactic equation in the lunar theory by Professor Hansen and the Astronomer Royal, Professor Airy ; 2nd, By the lunar equation in the theory of the Earth's motions, investigated by M. LeVerrier ; 3rd, By the excessive motions of Venus's nodes, and of the perihelion of Mars, also investigated by the same distinguished astronomer ; 4th, By the velocity of light, which is 183,470 miles per second, being a decrease of nearly 8,000 miles ; and 5th, By the observations on Mars during the oppositions of 1860 and 1862.

A diminution in the Sun's distance will necessarily involve a corresponding change in the masses and diameters of the bodies composing the Solar system. The Earth's mass will require an increase of about one-tenth part of the whole.

Substituting LeVerrier's solar parallax ($8''.95$) in Eq. (38), the Earth's mean distance from the Sun becomes 91,333,670 which is a reduction of 4,048,800 miles. The Sun's apparent diameter at the Earth's mean distance $= 32' \, 3''.64$, and in order that a body may subtend this angle, at a distance of 91,333,670 miles, it must have a diameter of 851,700 miles, which is a diminution of 37,800 miles. The distances, diameters, and velocities of all the planets in our system will require corresponding corrections if we express them in miles. Since the periodic times of the planets are known with great precision, we can easily determine by Kepler's third law, their mean distance from the Sun in terms of the Earth's mean distance. Thus: if T and t be the periodic times of the Earth and a planet respectively, and D the planet's mean distance, then regarding the Earth's mean distance as unity, we have $T^{\frac{2}{3}} : t^{\frac{2}{3}} :: 1 : D$

Or, $$D = \left(\frac{t}{T}\right)^{\frac{2}{3}}, \qquad (39).$$

In the case of Neptune the mean distance is diminished by about 121,000,000 miles. Jupiter's mean distance is diminished 21,063,000 miles, and his diameter becomes 88,296 miles, which is a decrease of 3,868 miles. These numbers shew the great importance which belongs to a correct knowledge of the Solar parallax.

(III.)

A TRANSIT OF MERCURY.

MAY 6TH, 1878.

Transits of Mercury occur more frequently than those of Venus by reason of the planet's greater velocity. The longitudes of Mercury's nodes are about 46° and 226°, and the Earth arrives at these points about the 10th of November and the 7th May, transits of this planet may therefore be expected at or near these dates, those at the ascending node in November, and at the descending node in May.

Mercury revolves round the Sun in 87.9693 days, and the Earth in 365.256 days. The converging fractions approximating

$$\text{to } \frac{87.9693}{365.256} \text{ are } \frac{7}{29}, \frac{13}{54}, \frac{33}{137}, \&c,$$

Therefore when a transit has occured at one node another may be expected after an interval of 13 or 33 years, at the end of which time Mercury and the Earth will occupy nearly the same position in the heavens.

Sometimes, however, transits occur at the same node at intervals of 7 years, and one at either node is generally preceded or followed by one at the other node, at an interval of $3\frac{1}{2}$ years.

The last transit at the descending node occurred in May, 1845, and the last at the ascending node in November, 1868. Hence the transits for the 19th century will occur, at the descending node May 6th, 1878 ; May 9th, 1891 ; and at the ascending node November 7th, 1881, and November 10th, 1894.

COMPUTATION OF THE TRANSIT OF 1878.

From the tables* of the planet we obtain the following heliocentric positions :—

---

* Tables of Mercury, by Joseph Winlock, Prof. Mathematics U. S. Navy, Washington, 1864.

Washington Mean Time.	Mercury's Helioc. Longitude.	Mercury's Helioc. Latitude.	Log. Rad. Vector.
1878, May 6d. 0h.	225° 52′ 57″.0	7′ 17″.3 N.	9,6545239
" 1h.	226 0 15 .4	6 23 .4	9,6546389
" 2h.	226 7 33 .6	5 29 .6	9,6547535
" 3h.	226 14 51 .6	4 35 .8	9,6548677

The following positions of the Earth have been obtained from Delambre's Solar Tables, corrected by myself, π being taken equal to 8″.95 at the Earth's mean distance :—

Washington Mean Time.	Earth's Helioc. Longitude.	Log. Earth's Rad. Vector.
1878, May 6d. 0h.	226° 0′ 38″.9	10,0040993
" 1h.	226 3 04 .0	10,0041038
" 2h.	226 5 29 .1	10,0041082
" 3h.	226 7 54 .2	10,0041126

The Sun's true longitude is found by subtracting 180° from the Earth's longitude.

Passing to the true geocentric places by Formulæ (3), (4), and (5), we obtain :—

Washington Mean Time.	Mercury's true Geoc. Longitude.	Mercury's true Geoc. Latitude.
1878, May 6d. 0h.	46° 6′ 52″.4	5′ 53″.6 N.
" 1h.	46 5 20 .4	5 10 .2
" 2h.	46 3 48 .3	4 26 .8
" 3h.	46 2 16 .3	3 43 .4

Formula (7) gives log. distance from Earth at 1h. = 9.7466455.

This will be required in formulæ (14) and (15) for finding the aberration.

Formula (9) gives $P = 15″.9$.

The semi-diameter of Mercury at the Earth's mean distance, $3″.34 = d'$, therefore by Eq. (12), $d = 5″.98$.

Aberration in Longitude = + 6″.67, by Eq. (14).

Aberration in Latitude = + 3″.34, by Eq. (15).

The Sun's semi-diameter = 15′ 52″.3. (Solar Tables).

The Sun's aberration = − 20″.25.

Correcting for aberration we obtain the apparent places as follows :—

Washington Mean Time.	Mercury's Appar. Geoc. Longitude.	Mercury's App. Geoc. Lat.	Sun's Appar. Longitude.
1878, May, 6d. 0h.	46° 6′ 59.″0	5′ 56.″9N.	46° 0′ 18″.7
" 1h.	46 5 27.0	5 13.5	46 2 43.8
" 2	46 3 54.9	4 30.1	46 5 8.9
" 3	46 2 26.9	3 46.7	46 7 34.0

Interpolating for the time of conjunction and collecting the elements, we have

Washington mean time of conjunction in longitude,
May 6d. 1h. 41 min. 17 sec.

Mercury's and Sun's longitude........... ... 46° 4′ 23″.6
Mercury's latitude..... .... 4′ 43″.6 N.
Sun's hourly motion in longitude .... 2′ 25″.1 E.
Mercury's hourly motion in longitude...... 1′ 32″.1 W.
Mercury's hourly motion in latitude......... 43″.4 S.
Sun's equatorial horizontal parallax......... 8″.87
Mercury's equatorial horizontal parallax ... 15″.9
Sun's semi-diameter..... .... 15′ 52″.3
Mercury's semi-diameter .... 5″.9

Employing the same notation as in Art. 13, the preceding elements give the following results. Relative hourly motion in longitude $= 3′ 57″.2$; $n = 10° 22′ 7″$; $m n = 241″.13$ the relative hourly motion in apparent orbit. $C F$ the least distance between the centres $= 279″$; $E F = 51″.04$; time of describing $E F = 12$ m. 42 sec. Since Mercury is *north* of the Sun's centre at conjunction, and moving southward, $E F$ will lie on the *right* of $C E$ (see *Fig.* 4), and the middle of the transit will take place at 1h. 54m. P.M.

Sum of semi-diameters $= 958″.2$

$$V = 16° 55′ 44″ ; \quad V F = 916″.68 ;$$

Time of describing $V F = 3$h. 48.1 min. $=$ half of the duration. Subtracting 3h. 48.1 min. from, and adding the same to the time of the middle of the transit, we obtain the times of the first and last contacts, as seen from the Earth's centre, thus :

First external contact May 6d. 10h. 5.9 min. A,M.

Last external contact   "   5h. 42.1 min. P.M.

Mean time at Washington,

The places which will have the Sun in the zenith at these times can be found in the same manner as in Art. 14, with the aid of the following elements :—

Obliquity of the Ecliptic 23° 27' 25".

Sidereal time at Washington at mean noon of May 6th (in arc) 44° 24' 50".46.

Since the relative parallax is only 7" the time of the first or last contact will not be much influenced by the parallax in longitude and latitude, and therefore the preceding times for Washington are sufficiently accurate for all ordinary purposes.

The mean local time of beginning or end for any other place, is found by applying the difference of longitude, as below :—

The longitude of Washington is 5h. 8m. 11 sec. W.

The longitude of Toronto   is 5h. 17m. 33 sec. W.

Therefore Toronto is 9 min. 22 sec. west of Washington.

Then, with reference to the centre of the Earth, we have for Toronto,

First external contact May 6d. 9h. 56.5m. A.M.

Last external contact   "   5h. 32.7m. P.M.

                                        Mean time.

For Quebec, longitude 4h. 44m. 48 sec. W.

First external contact May 6d. 10h. 29.3m. A.M.

Last external contact   "   6h. 6.5m. P.M.

                                        Mean time.

For Acadia College, longitude 4h. 17.6m. W.

First external contact May 6d. 10h. 56.5m. A.M.

Last external contact   "   6h. 32.7m. P.M.

                                        Mean time.

For Middlebury College, Vermont, longitude 4h. 52.5m. W.

First external contact, May 6h. 10h. 21.5m. A.M.

Last external contact   "   5h. 57.7m. P.M.

                                Mean time at Middlebury.

APPENDIX.

Eclipses of the Sun are computed in precisely the same way as transits of Venus or Mercury, the Moon taking the place of the planet. The Solar and Lunar Tables furnish the longitude, latitude, equatorial parallax, and semi-diameter of the Sun and Moon, while Formulæ (19)–(30) furnish the parallax in longitude and latitude. If the computation be made from an ephemeris which gives the right ascension and declination of the Sun and Moon instead of their longitude and latitude, we can dispense with formulæ (21) and (23), and adapt (25), (26), (29), and (30) to the computation of the parallax in right ascension and declination. In *Fig.* 6, let Q be the pole of the equator, then $L\,Q$ is the co-latitude $= 90° - \phi$; $Z\,Q\,S = h$, the Moon's true hour angle $=$ the Moon's A. R. $-$ the sidereal time; $S\,Q\,S'$ is the parallax in A. R. $= x$, and $Q\,S' - Q\,S$ is the parallax in declination $= y$. Put $Q\,S$, the Moon's true north polar distance $= 90 - \delta$, then Formulæ (25) and (26) become,

$$\sin x = \sin P \cos \phi \sec \delta \sin (h + x) \qquad (25, \text{bis}).$$
$$= k \sin (h + x)$$

$$\text{Or,} \quad x = \frac{k \sin h}{\sin 1''} + \frac{k^2 \sin 2h}{\sin 2''} + \frac{k^3 \sin 3h}{\sin 3''} + \&c. \qquad (26, \text{bis}).$$

Again, the formulæ for determining the auxiliary angle θ in (29) becomes,

$$\cot \theta = \cot \phi \cos (h + \tfrac{x}{2}) \sec \tfrac{x}{2}.$$

And (29) becomes,

$$\sin y = \sin P \sin \phi \operatorname{cosec} \theta \sin ((\theta - \delta) + y). \qquad (29, \text{bis}).$$
$$= k \sin ((\theta - \delta) + y)$$

$$y = \frac{k \sin (\theta - \delta)}{\sin 1''} + \frac{k^2 \sin 2 (\theta - \delta)}{\sin 2''} + \frac{k^3 \sin 3 (\theta - \delta)}{\sin 3''} + \&c.$$

(30, bis).

These parallaxes when applied with their proper signs to the right ascensions and declinations of the Moon for the assumed times, furnish the *apparent* right ascensions and declinations. The difference between the *apparent A. R.* of the Moon and the *true A. R.* of the Sun, must be reduced to seconds of *arc* of a *great circle*, by multiplying it by the cosine of the Moon's *apparent* declination. The apparent places of the Moon with respect to the Sun will give the Moon's apparent orbit, and the times of apparent contact of limbs are found in the same way as described in Art. 13. The only other correction necessary to take into account, is that for the augmentation of the Moon's semi-diameter, due to its altitude. The augmentation may be taken from a table prepared for that purpose, and which is to be found in all good works on Practical Astronomy, or it may, in the case of solar eclipses, be computed by the following formulæ :—

TO FIND THE AUGMENTATION OF THE MOON'S SEMI-DIAMETER.

Let C and M be the centres of the Earth and Moon, A a point on the Earth's surface, join $C M$, $A M$, and produce $C A$ to Z; then $M C Z$ is the Moon's true zenith distance $= Z =$ arc $Z S$ in *Fig. 6*; and $M A Z$ is the apparent zenith distance $= Z' =$ arc $Z S'$ in the same figure. Represent the Moon's semi-diameter as seen from C, by d; the semi-diameter as seen from A by d'; the apparent hour angle $Z Q S'$ by h', and the apparent declination by δ', then

$$\frac{d'}{d} = \frac{C M}{A M} = \frac{\sin Z'}{\sin Z}$$

$$= \frac{\sin Z S'}{\sin Z S} \quad (\text{See Fig. 6.}) \qquad (40).$$

$$= \frac{\sin h' \cos \delta'}{\sin h \cos \delta} , \text{ by Art. 21.}$$

Therefore, $\qquad d' = d . \dfrac{\sin h' \cos \delta'}{\sin h \cos \delta} ,$ $\qquad\qquad$ (41).

This formula furnishes the augmented semi-diameter at once. It can be easily modified so as to give the augmentation directly, but with logarithms to seven decimal places, it gives the apparent semi-diameter with great precision.

As examples we give the following, the first of which is from Loomis's Practical Astronomy :—

Ex. 1. Find the Moon's parallax in *A. R.* and declination, and the augmented semi-diameter for Philadelphia, Lat. 39° 57′ 7″ N. when the horizontal parallax of the place is 59′ 36″.8, the Moon's declination 24° 5′ 11″.6 N., the Moon's true hour angle 61° 10′ 47″.4, and the semi-diameter 16′ 16″.

Ans.—Parallax in *A. R.*, 44′ 17″.09
" Dec., 26′ 10″.1
Augmented semi-diam = 16′ 26″.15.

Ex. 2. Required the times of beginning and end of the Solar Eclipse of October 9–10, 1874, for Edinburgh, Lat. 55° 57′ 23″ N. Long. 12m. 43 sec. West, from the following elements obtained from the English Nautical Almanac :—

Greenwich mean time of conjunction in *A R*,

	Oct. 9d.	22h.	10m.	11.4 sec.
Sun's and Moon's *A R*		195°	36′	30″
Moon's declination S		5	39	8.9
Sun's declination S		6	39	34.1
Moon's hourly motion in *A R*			26	21.9
Sun's do			2	18.2
Moon's hourly motion in Declination. S			13	48.3
Sun's do S				56.9
Moon's Equatorial Horizontal Parallax.			53	59.6
Sun's do do				9.0
Moon's true semi-diameter			14	44.2
Sun's do			16	3.8
Greenwich sidereal time at conjunction.		171	23	32.8

Assuming, for the beginning, 20h. 55m., and for the end, 23h. 10m. Greenwich mean time, we obtain from the preceding elements and formulæ the following results, which may be verified by the Student :—

Geocentric latitude = 55° 46′ 41″ ; reduced or relative Parallax = 53′ 43″.2.

	20h. 55m. G. M. T.	23h. 10m. G. M T.
Moon's $A R$	195° 3′ 27″.6	196° 2′ 46″9
Sun's $A R$.	195 33 36.8	195 38 47.8
Moon's Dec	5 21 50.9 S.	5 52 54.6 S.
Sun's Dec	6 38 22.9 S.	6 40 30.9 S.
Sid. Time at Edin. (in arc).	149 21 51.5	183 12 24.1
Moon's true hour angle...	45 41 36.1 E.	12 50 22.8 E.
Moon's Parallax in $A.R$...	+ 21 49.4	+ 6 48.5
Moon's do in Dec...	+ 46 25.1	+ 47 32.7
Moon's apparent $A R$	195 25 17.0	196 9 35.4
Moon's do Dec	6 8 16.0 S.	6 40 27.3 S.
Diff. of $A R$ in seconds of arc of great circle	496″·9, Moon W.	1835″.1 Moon E.
Diff. Dec	30′ 6″.9, Moon N.	3″.6 Moon N.
Aug. semi-diam of Moon...	888″.4	890″.5.

Eclipse begins October 10d. 8h. 43m. 32 sec. A.M.

Eclipse ends   "   10h. 58m. 22 sec. A.M.

Mean time, at Edinburgh. Magnitude .369 Sun's diam.

THE END.